# Neutron Charge

by

Patrick John Naughton

Patrick Naughton

**Dedicated to Kathy.**

Published by www.siliconpool.com

ISBN 978-0-244-92345-7

Printed and bound in the United Kingdom by www.lulu.com

Version 2
(SS 2017A)

# Contents

Patrick Naughton

# Neutron Charge

**Abstract**

Here in this script I propose a novel method for calculating the charge of a neutron. Current prevailing Physics claims that the size of charge associated with a neutron is zero. However according to my calculations there might be a small charge associated with a neutron and the value of this charge could be;-

$$\mathbf{Q_N} = \mathbf{+4.7 \times 10^{-23} \pm 0.1 \times 10^{-23}\ Coulombs}$$

This might also be the size of the charge related to the neutrino although in the case of the neutrino the charge is likely to be an opposite negative charge of **-4.7 x 10$^{-23}$ ± 0.1 x 10$^{-23}$ Coulombs**.
If this is correct then we would need about 3,409 neutrinos to equal the magnitude of charge of an electron assuming the currently accepted charge on an electron is correct at a value of;

$$e = -1.6021927 \times 10^{-19}\ Coulombs$$

**Patrick Naughton August 2017**

Neutron Charge

## Introduction

I have been an Applied Physicist at heart since as long as I can remember. Therefore it is part of my DNA to always look for the application in any possible new science. Of course it isn't only the duty of the Applied scientists to look for the ways in which new discoveries can be leveraged for the purposes of advancing the interests of the world at large, but it is certainly something that we have been schooled to consider. However I have decided to publish this work before I have concluded the process of analysing the possible accuracy or application of these proposals.

The model that I propose should not be taken in any way as a concrete claim of absolute truth. It has *not* been rigorously tested in a way that would be required to establish any such claim. This work should be considered as no more than a proposal at this point in time. It has certainly not been confirmed or validated in any way. However it is hoped that these ideas may help others to establish improved models of the dynamics of electron orbits, and it is hoped they may at least inspire others to derive better proposals, that will in some way bring benefit to the world at large.

I make no apology for bringing my ideas 'to market' too soon. I felt I had no real choice. If I had waited until I had tested every angle of this theory, I would have kept the world waiting a very long time. I have neither the resources nor the energy to undertake such a task. Better I think that I have put my ideas out there for others to correct and possibly build on as they see fit, or discredit and reject whichever might be the most appropriate course of action.

At the present time we have two dominant theories in Physics that are both very successful in terms of predicting observations - these are Quantum Mechanics and Einstein's theory of General Relativity. Quantum Mechanics is extremely accurate at predicting the behaviour of very small particles whilst General Relativity is equally as good at predicting the behaviour of large scale bodies such as planets. However in many ways these two important theories seem quite incompatible. I decided to apply the theory of General Relativity to atoms in the hope that this approach might lend something to the ongoing quest to bring together these two great theories.

I do believe what I've written here might have some merit and may be of some help towards a better understanding of the world we live in but that remains to be seen.

This is the result of work I have produced myself with no collaboration with anyone else of any kind. As such, many people may choose to dismiss my findings as pure conjecture. They may well be right. This isolated, individualistic approach is certainly not the best approach and certainly not the one that I would recommend for the purposes of conducting scientific endeavour, as this approach certainly constrains progress. However it was the only realistic option open to me and therefore the approach that I simply had to follow as the pressures of paid work had, by necessity, to take priority. I have written this short speculative paper with a view to gaining feedback. As such nothing that I've written here should be taken as proven nor should any of this theory be relied on for any purposes whatsoever.

I realise this paper is not written in the normal style of scientific papers. I hope this doesn't detract from the overall result that I am hoping to communicate. I made a conscious decision to make this paper 'less formal' with the hope that this might make the proposals that I'm trying to put over perhaps a little bit more accessible. It's difficult to gauge how this might work out, but I can assure you it has been done with the very best of intention.

Finally I have to admit to having more than one motive in the creation of this work. I have huge admiration for all people involved in constructive scientific endeavour, and I am constantly impressed by the energy and enthusiasm of young people engaged in scientific discovery. But one thing that particularly annoys me is the widely held assumption that no one who has achieved the age of fifty years can ever have anything meaningful to contribute to the debate. There is a widespread but mistaken belief that if anyone had had anything useful to contribute they would have done so long before their fiftieth year. I am well over fifty years old and I am quite determined to prove that this unfounded bias is completely wrong. I realise many scientists such as Einstein and Newton started achieving great successes at very young ages, but I'd like to prove that age is no barrier to success, and that meaningful achievements can also come from those of us with more experience as well as from our younger colleagues.

## Disclaimer /Health Warning

Please note that this work, along with many of my writing on science is mostly my own scientific theory. As such these works should not be accepted at face value as undisputable scientific fact. My thoughts in this area are little more than suggestions that are largely unproven and in many cases are quite possibly or very likely to be incorrect. I do not set out to write science fiction but it is quite possible that I might often end up doing so. This work should be thought of as *scientific possibility*. A very great man once said the best thing you could do for anyone was to inspire them. I try to set down some of my original thoughts in writing. If I inspire anyone to do anything meaningful than I will consider myself a success. Writing in this way gives me a freedom to cover more ground than I'd ever achieve if I were to constrain myself to a strict fundamental, scientific approach. I realise this is not the normal approach to science and as such it is one that many will resent, but I believe it will ultimately provide me with the most effective vehicle to any success.

Currently the prevailing conventional theory claims that electrons do not actually orbit as hard spherical particles around the nucleus of an atom but are in fact more akin to clouds of energy. I've chosen to use the 'hard ball' model of electrons in this calculation, merely for the sake of simplicity, without any claim that this is any way necessarily accurate, but at least this model lends itself to more simple, straight forward calculation, albeit one that may not necessarily be in any way accurate.

### Einstein Kinetic Energy

According to Einstein the Kinetic Energy of a particle is given by the expression;-

$$KE = mc^2(Y - 1) \qquad \qquad ...1$$

where m is the rest mass of the particle and c is the speed of light and Y is the Lorentz coefficient which itself can be expressed as;-

$$Y = c/(c^2 - V^2)^{0.5} \qquad \qquad ...2$$

where V is the speed of the particle.

The equation for Y mandates that the maximum relative speed a particle can reach is 'c' the speed of light. According to Einstein the speed of light is in effect the Universal speed limit. There does seem to be a problem with this definition. We are turning here on the surface of the Earth at a rate of about 360° every 24 hours or thereabouts. That means that anything more than about 2.8 billion miles away is moving relative to us at a speed greater than the speed of light, which includes every single star other than our own Sun. Our nearest star is Alpha Centauri which is about 25 trillion miles away. However for the purposes of this calculation I will accept Einstein's definition of Kinetic Energy.

The Universe is made up of many different elements which are characterised by atoms which themselves, in their basic stable state, contain different numbers of electrons, protons and neutrons. Electrons can of course be removed from atoms by ionisation. If an electron receives the right amount of energy it can escape from an atom. I decided to study the *last remaining electron* in each atom in an attempt to understand more about the way electrons orbit within their host atoms. The main motive behind choosing to study the last electron in each atom was to keep the calculations involved as simple as possible.

Only certain electron orbits and therefore only corresponding energy levels are allowed in an atom. This is a central tenet of Quantum Theory. For an electron to complete an orbit perpetually within an atom, it must achieve a certain harmony. Just like the plucking of a guitar string must achieve either a half wavelength or a full wavelength, then similar constraints must exist within an atom.

Diagram 1 below shows the first harmonic that is achieved when you pluck a guitar string. The length of the string is half a wavelength for the wave in the vibrating string.

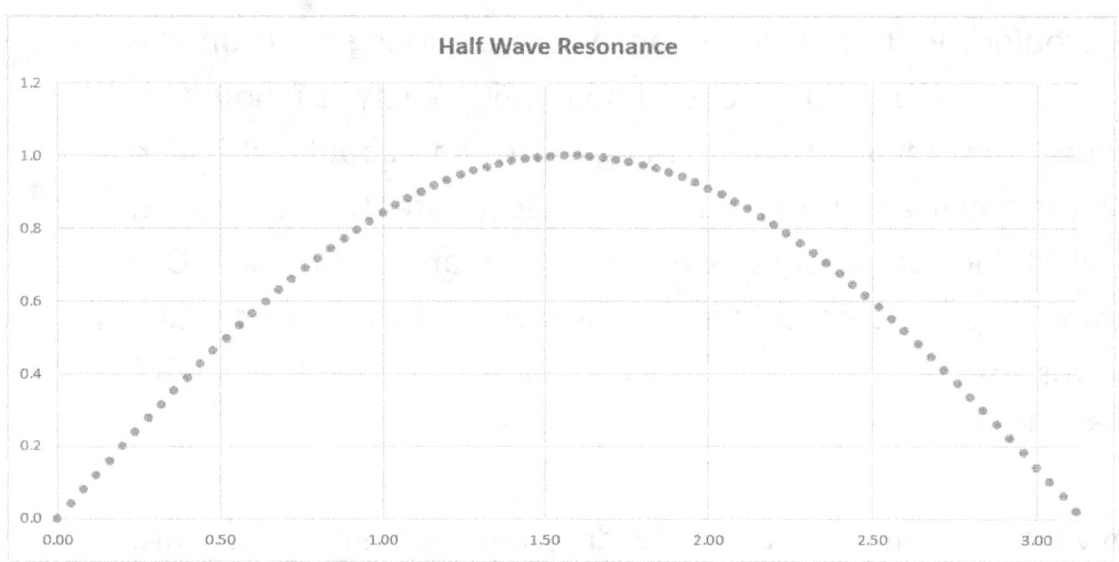

Diagram 1

A next harmonic is achieved when the frequency of the vibration is doubled and the wavelength of vibration in the string is halved as shown in diagram 2 below;-

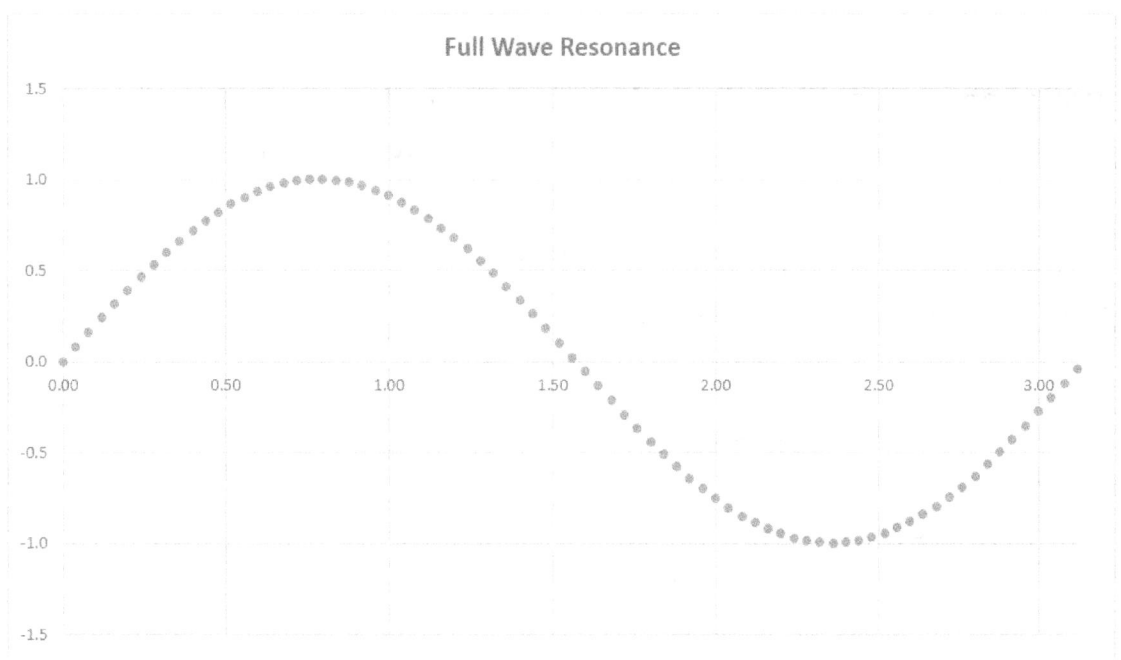

Diagram 2

An electron is said to have a spin of S = 1/2. It is perhaps more convenient to speak in terms of the reciprocal spin S' where S' = 1/S which means for an electron S' = 2.

In practical terms at a fundamental level this means that an electron must complete two orbits within an atom to maintain a stable orbit. A simple analogy of this can be demonstrated by imaging a playing card spinning as it completes a circular orbit around a clock face. Imagine the 'Queen of Hearts' starting an orbit face up in the 12 o' clock position. Imagine that after a complete orbit that the card has only spun half a turn so now back at the 12 o' clock position the Queen is facing downwards. At this rate of spin (S = 1/2) the Queen has to complete a second orbit to return to her original position facing upwards again at the 12 o' clock position.

This same process prevails for an electron orbiting in an atom.

## Proton Attraction

The amount of energy required to ionize the last electron out of each atom is well known. Values are listed in appendix 2. From these values we can use equation 1 to calculate the values of Y the Lorentz coefficient for each of the last electrons. Plotting these values against the number of protons P in each atom gives us graph1 shown below.

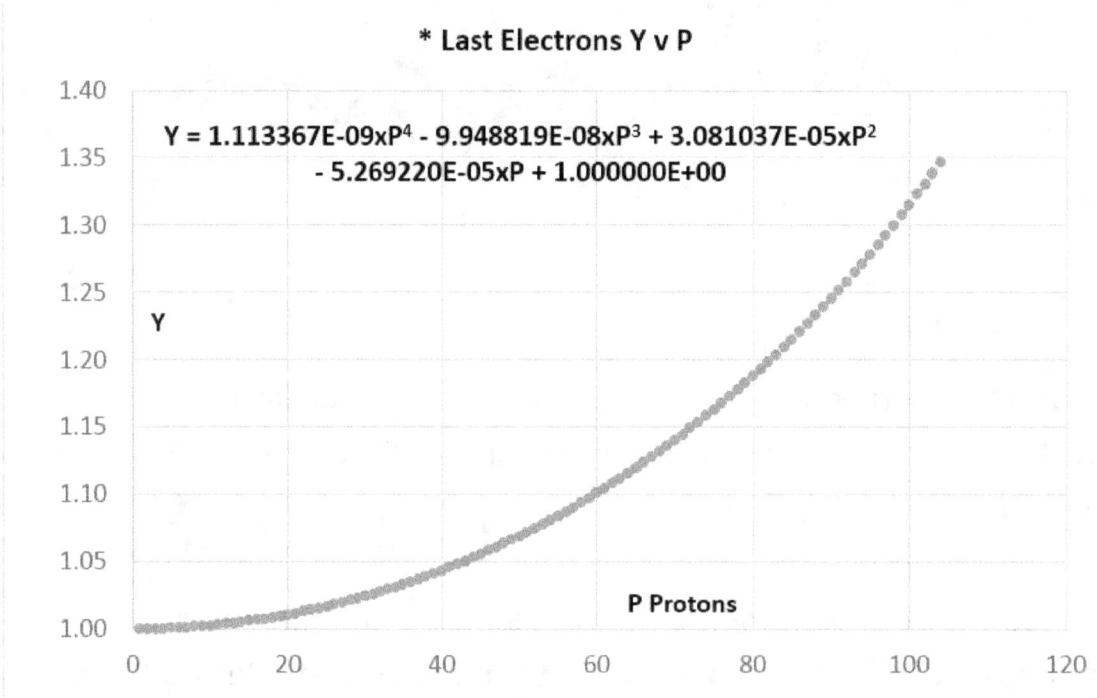

Graph 1

Graph 2 below shows the ionisation energy plotted against P and as we might expect it is a similar shaped curve.

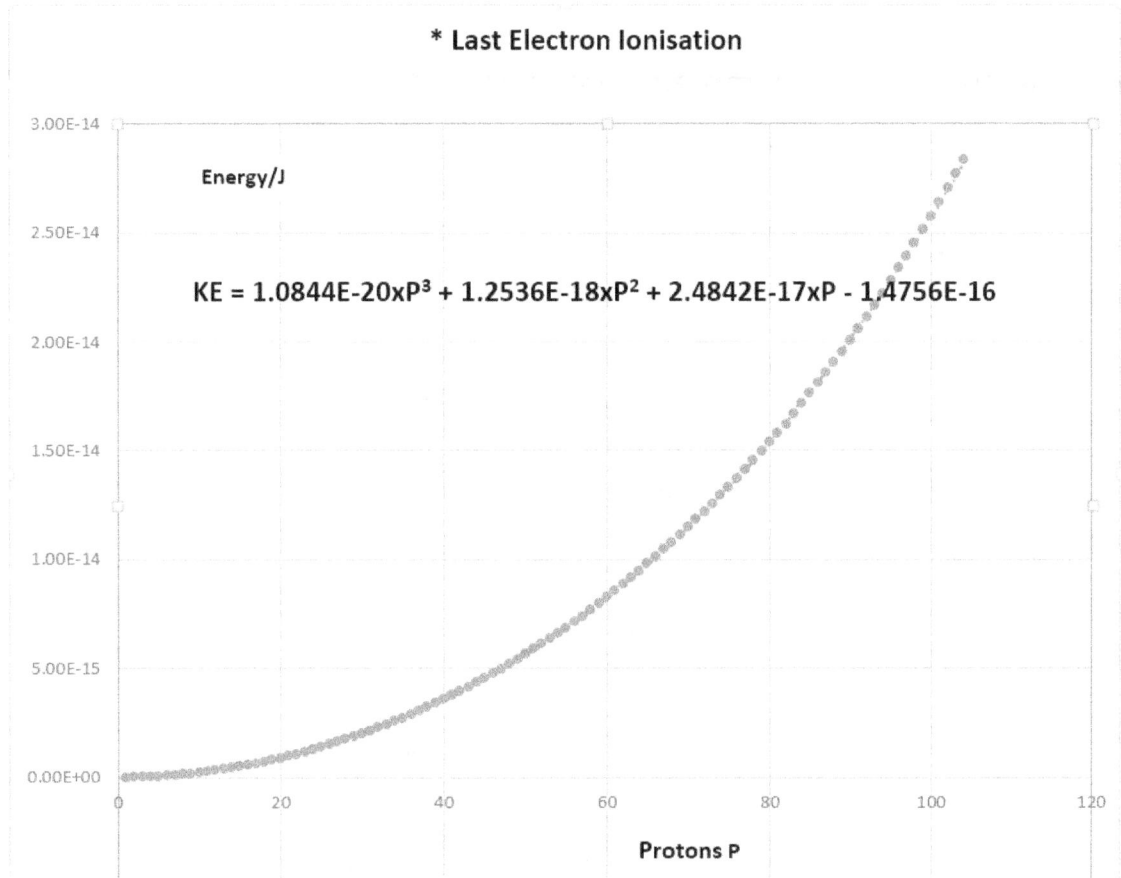

**\* Last Electron Ionisation**

Energy/J

$$KE = 1.0844E{-}20 \times P^3 + 1.2536E{-}18 \times P^2 + 2.4842E{-}17 \times P - 1.4756E{-}16$$

Protons P

Graph 2

## Frequency Matched Ionization

The frequency of the light contained in the ionizing photon must match the frequency of the electron spinning around the atom in some way in order for ionisation to occur. The frequency of an electron spinning in a circular orbit around an atomic nucleus i.e. the number of times it orbits the atom per second, is given by;-

$$f_{electron} = V/2\pi R \qquad\qquad\qquad ...3$$

where V is the velocity of the electron and R is the radius of the circular orbit.

11

The reciprocal spin of an electron is nominally given as S' = 2. This means the electron has to complete <u>two</u> orbits of an atom to get back to the same position, in the same state as it originally started.

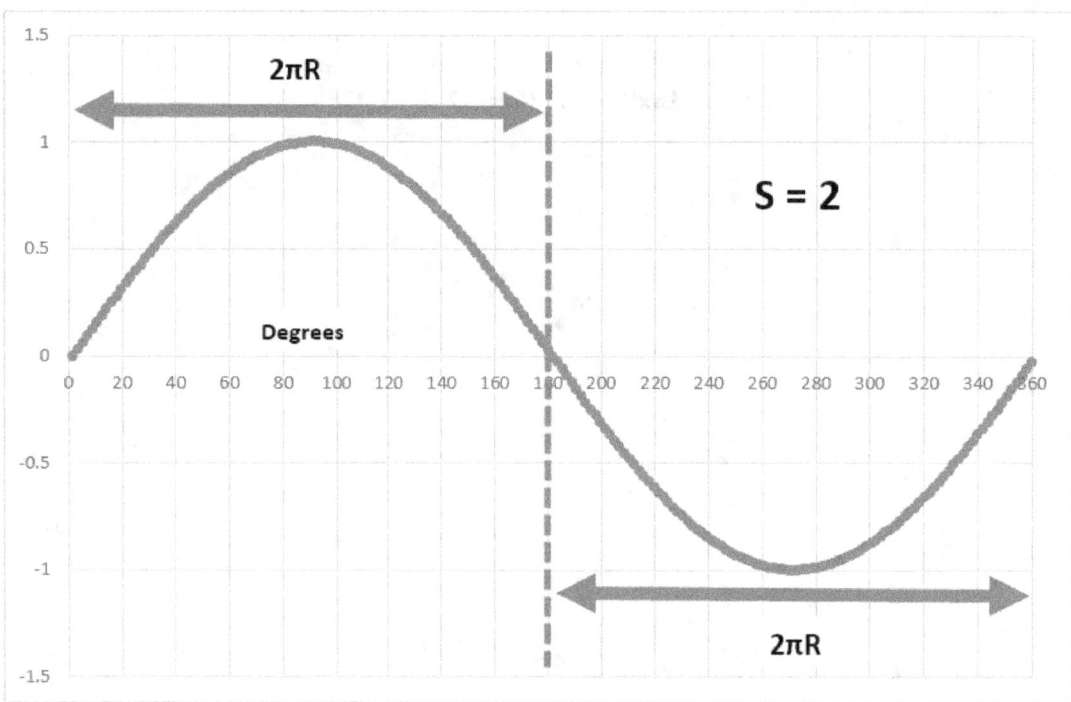

Diagram 3

We can surmise that for frequency matching to occur that;-

$$f_{ion} = f_{electron} \qquad \qquad \qquad ...4$$

where $f_{electron}$ is the deBroglie frequency of the motion of the electron in its two orbit cycle around its atom (the frequency at which it completes two orbits) and $f_{ion}$ is the Planck frequency of the photon that causes ionisation.

However if we analyse the attraction between the last electron in any atom to the protons in the nucleus of that atom using Coulomb's law alone to calculate only the force of attraction between the last remaining electron in an atom and its nucleus, then we find that the

Coulomb force is not sufficient to account for the spin of the electron to achieve the required spin S' = 2

According to General Relativity the force F required to achieve circular motion for a mass m travelling at velocity v in a circle of radius r is given by;-

$$F = Y^2 \, mv^2/r \qquad\qquad ...5$$

where Y is the Lorentz coefficient.

According to Coulomb's Law the attraction between the last electron in an atom and the P protons in the nucleus of the atom is given by;-

$$F = Y^2 Pe^2/4\pi\varepsilon r^2 \qquad\qquad ...6$$

where $\varepsilon$ is the permittivity of free space and e is the charge of the electron and the charge of each proton. (Please note: Y is the Lorentz coefficient and is not something that Coulomb included, and you may wonder whether it is correct to include it here. However at low speed it is important that the results predicted by General Relativity agree with the Classical theory and if we did not include the $Y^2$ in Coulomb's law we would not get agreement that at low *circular* velocities v that Potential Energy PE is related to Kinetic Energy by the expression;-

$$PE = -2 \times KE \qquad\qquad ...7$$

This is explained further in appendix 1.

To get S' to be constant equal to 2 then we need to introduce another force $F_1$.

So let us say that the total force $F_T$ between the last electron and the nucleus in the atom is given by;-

$$F_T = y^2 mv^2/r = y^2(Pe^2/4\pi\varepsilon r^2 + F_1) \qquad \ldots 8$$

From the energy required to ionize the last electron Eion out of each atom we can calculate the Lorentz value Y as explained earlier. The ionisation energy Eion must match the Kinetic Energy of the electron in the atom for their frequencies to match, which is the reason the energy of the photon is absorbed by an electron.

So we can say that KE = Eion (which I'll refer to simply as E).

We also know from DeBroglie and others the relationships required to calculate the effective radius of the atom;-

$$E = hf = hv/\lambda \qquad \ldots 9$$

where E is the Kinetic Energy of the particle and v is its velocity and $\lambda$ is the deBroglie wavelength.

We know that;-

$$\lambda = 2\pi r \times S' \qquad \ldots 10$$

and since S' = 2 (assuming the electron spin is 1/2) then we can say;-

$$E = hv/4\pi r \qquad \ldots 11$$

$$y^2 = c^2/(c^2 - v^2) \qquad \ldots 12$$

so

$$c^2 - v^2 = c^2/y^2 \qquad \qquad \ldots 13$$

Therefore...

$$v = (c/y) \times (y^2 - 1)^{0.5} \qquad \qquad \ldots 14$$

which means...

$$E = hc \times (y^2 - 1)^{0.5} / 4\pi r y \qquad \qquad \ldots 15$$

$$r = hc \times (y^2 - 1)^{0.5} / 4\pi E y \qquad \qquad \ldots 16$$

Having been able to derive y and r and v from the ionisation energy Eion then we can easily calculate the additional force F1 required to produce a harmonic orbit for the last electron in any atom. The results are shown in the graph below;-

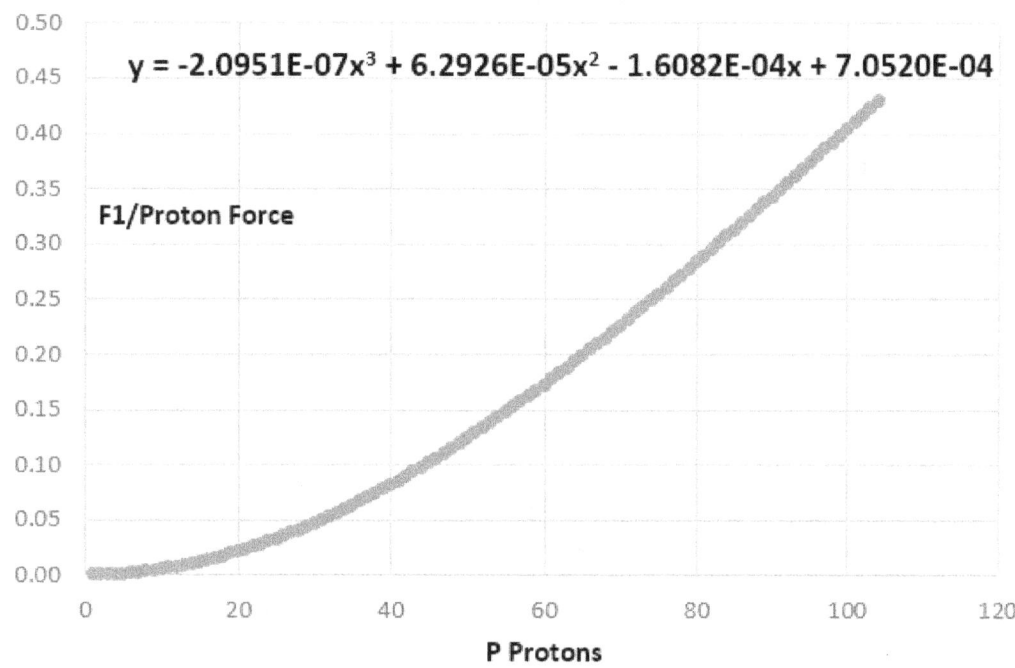

**Added Force/Proton Force Last Electron**

$y = -2.0951E-07x^3 + 6.2926E-05x^2 - 1.6082E-04x + 7.0520E-04$

F1/Proton Force

P Protons

Graph 3

It is possible that the neutrons in the nucleus of each atom could be responsible for the additional force.

We can say the additional force $F_1$ could be expressed by the relationship;

$$F_1 = P^W e \times Q_T/4\pi\epsilon r^2 \qquad \qquad ...17$$

where P is the number of Protons (and for many atoms the number of neutrons) in the nucleus and e is the charge of the last electron and $Q_T$ is the additional charge from the nucleus. The power W is raised on the value P for simplicity as this is the only dimensionless value in the equation and we need to preserve the dimensional balance of the relationship on both sides. If we take the natural logs of each side we get;-

$$\log_e(F_1 \times 4\pi\epsilon r^2) = W \times \log_e P + \log_e (e \times Q_T) \qquad ...18$$

If we plot $\log_e(F_1 \times 4\pi\epsilon r^2)$ against $\log_e P$ we get the straight line expected (see Graph 4 below) with a y intercept k where;-

$$k = \log_e (e \times Q_T) \qquad \qquad ...19$$

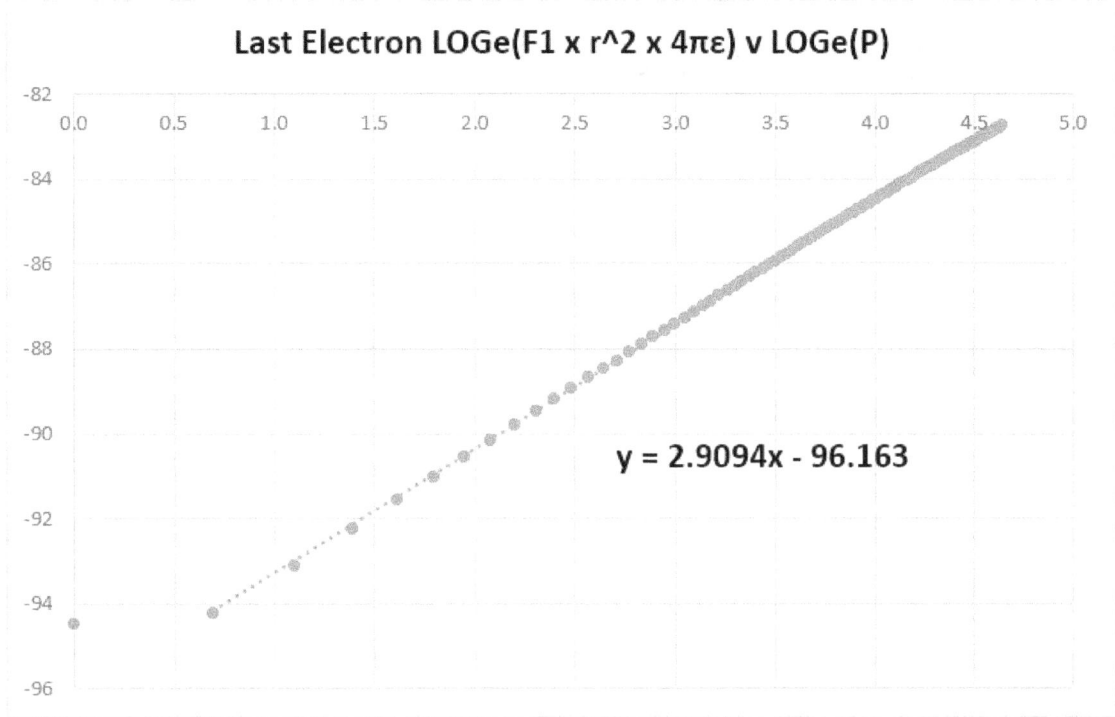

Graph 4

Everything looks as we might expect it to be in graph 4 except the lowest value, the value for hydrogen, is out of line.

This discrepancy might be explained by the lack of any neutron in the hydrogen atom. So if we look at the Y intercept that we'd expect if hydrogen did have a neutron, and the one we actually get for the hydrogen atom without any neutron, we can calculate the possible charge on a neutron.

$$K_2 = \log_e (Q_{T2} \times e) = -94.491 \text{ (proton only)}$$
$$K_1 = \log_e (Q_{T1} \times e) = -96.163 \text{ (proton \& neutron)}$$

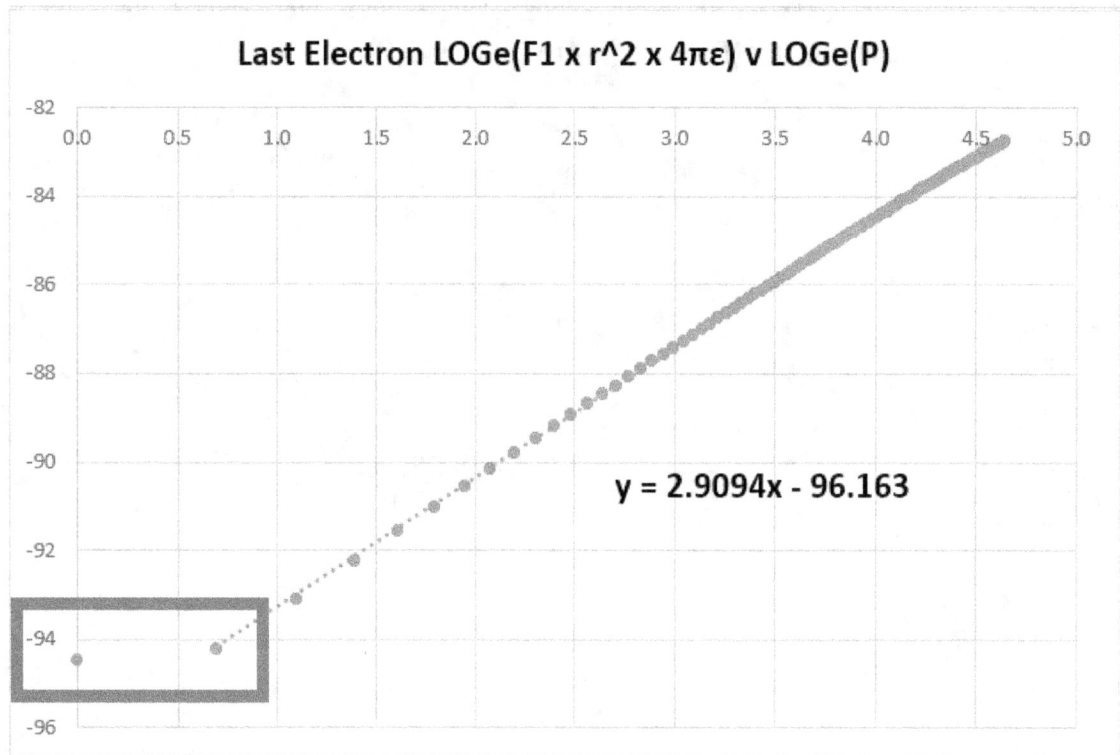

Graph 5

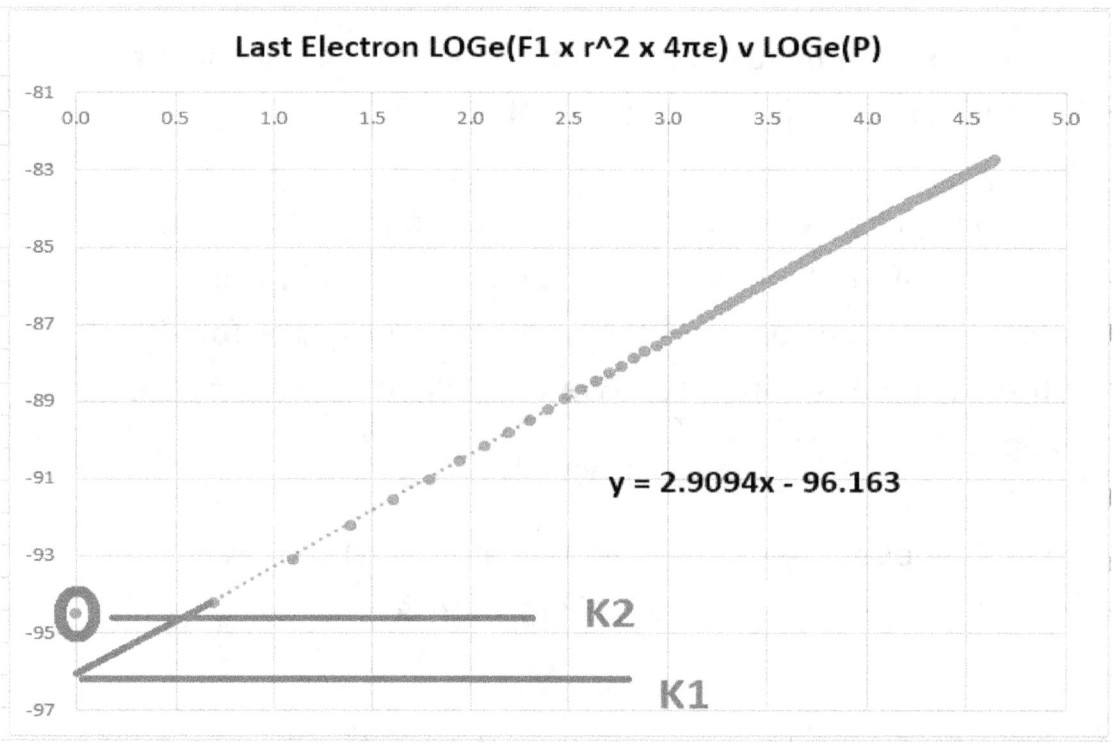

Graph 6

$$Q_{T2} = -1.0770 \times 10^{-23}$$

$$Q_{T1} = -5.7309 \times 10^{-23}$$

The value $Q_{T2}$ - $Q_{T1}$ = $Q_N$ which we can say is the charge which the neutron may be responsible for.

Calculating this out we find;-

**$Q_N$ = +4.65 x $10^{-23}$ Coulombs   ... (Result 1)**

We can use the value $Q_{T2}$ for the portion of charge in $F_1$ caused by a *single proton only* and equation 18 to calculate the value $Q_N$ (the charge caused by a single neutron) for each and every element. The total charge $Q_T$ which gives rise to the force $F_1$ is shown in the graph below;-

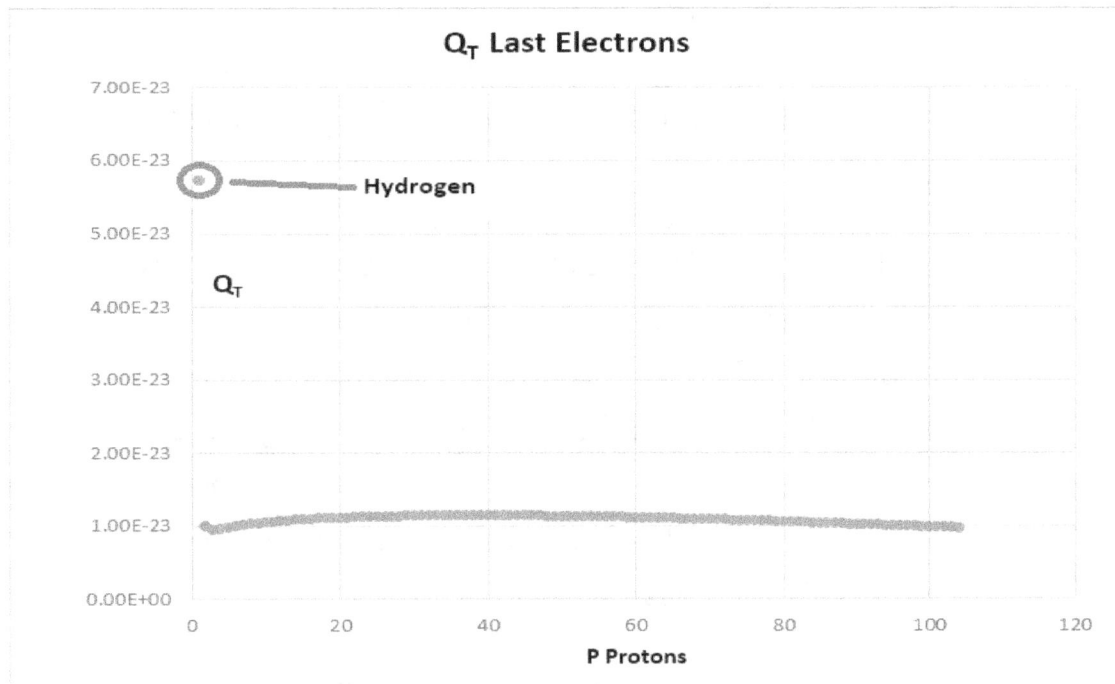

Graph 7

Ignoring the special case of the hydrogen atom we can see that for all the other elements the value of the charge on each neutron is approximately the same;-

$Q_T$ MIN = 9.3743 x$10^{-24}$

$Q_T$ MAX = 1.1388 x $10^{-23}$

$Q_T$ AV = 1.038 x $10^{-23}$ +/- 1.007 x $10^{-24}$

From this, and from the value of charge variant not caused by any neutron, as illustrated by the hydrogen atom, we can deduce the value of $Q_N$ the charge caused by each individual neutron is;-

**$Q_N$ = 4.7 x $10^{-23}$ ± 0.1 x $10^{-23}$**                    **... (Result 2)**

This is consistent with Result 1 that was derived earlier from the consideration of the hydrogen atom only. If this is correct it means that the charge on an electron is about 3,409 times bigger than the charge of the neutron and of opposite sign. That is to say the charge on the neutron is positive.

It is widely accepted that the fusion process in the Sun is driven by the collision of protons. When two protons collide in the core of the Sun it is believed that one of them breaks up into a neutron, a neutrino and a positron. If the positron has an identical charge to the original proton, and if charge is conserved throughout the process, then the neutron and neutrino are likely to have equal and opposite charges so the neutrino is likely to have a negative charge of;

**$Q_v$ = -4.7 x $10^{-23}$ ± 0.1 x $10^{-23}$**                    **... (Result 3)**

One point that remains unexplained from this calculation is the nature of the relationship of the force additional to the Coulomb force.

I have derived that the force $F_T$ between the last electron in an atom and the nucleus of the atom can be represented by;-

$$F_T = Y^2\{Pe^2/4\pi\varepsilon r2 + P^We \times Q_T/4\pi\varepsilon r^2\} \qquad \text{...20}$$

where from Graph 4 we can say W = 2.9094

This seems a somewhat strange result and at the time of writing is totally unexplained. Why should neutrinos (if that is indeed the source) multiply at a power that is nearly, but not quite, the cubed value of the number of protons present in the nucleus?

The number of neutrons N in each atom is related to the number of protons P as shown in graph 8. The relationship is;-

$$N = 0.0048 \times P^2 + 1.1466 \times P - 1.5863 \qquad \text{...21}$$

There is no obvious relationship to $P^{2.9094}$ that is seen in equation 20.

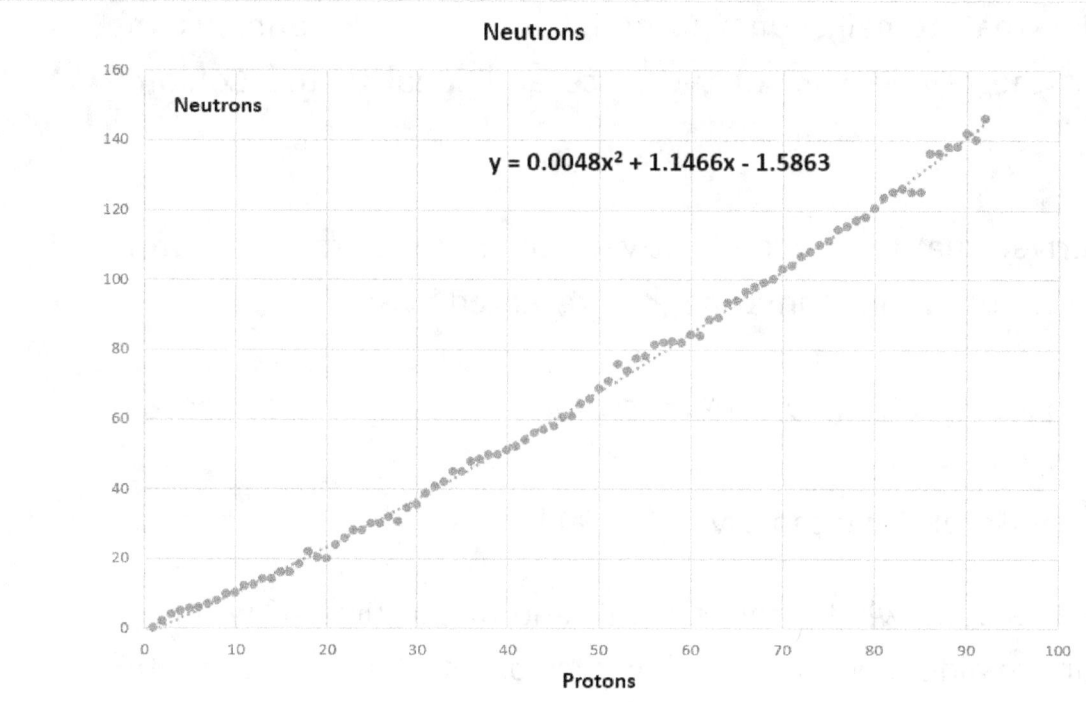

Graph 8

If the ratio of charge to mass q/m is the same ratio in a neutrino as it is in an electron (e/m = 1.7588378 x $10^{11}$ Coulombs/Kg for an electron) then we can estimate that the mass of a neutrino would be;-

$$M_N = 2.647065 \times 10^{-34} \text{ kg}$$

This is about $2.933 \times 10^{-4}$ x mass of electron

Actually so far the best experimental evidence suggests neutrinos have a mass that is about;-

$$M_N \sim 10^{-5} \text{ x mass of electron}$$

So q/m for the neutrino is probably about 29 times bigger than the charge to mass ratio of the electron or thereabouts.

The Sun produces neutrinos at an astonishing rate. It is thought to generate $1.72 \times 10^{38}$ neutrinos per second. At an estimated mass of $2.647065 \times 10^{-34}$ kg for each neutrino then there would be about 45,552 kg of neutrinos emitted by the Sun each second. With a Sun mass of $1.98841993902935 \times 10^{30}$ it would take about $1.38 \times 10^{18}$ years for the total mass of the Sun to evaporate into neutrinos. The Universe is thought to be about $13.9 \times 10^{9}$ years old so, at the present rate of production, the Sun should still have a long way to go before it has emitted all of its mass as neutrinos. At the present rate of emission the Sun would only have emitted about 0.000001% of its mass as neutrinos in a period of 13.9 billion years whereas it would have lost about 0.09% of its mass due to the energy of light photons it would have emitted in this period, which is about 100,000 times more than the loss of mass by neutrino emission. (In actual fact the Sun is thought to be only about 4.6 billion years old*1 so it has probably only lost about one third of these amounts of mass in its lifetime so far).

## Detecting And Measuring Neutrinos

When a particle falls from the sky towards the surface of the planet it loses Potential Energy which is converted to Kinetic Energy. That is to say that the particle speeds up as it falls downwards. According to Isaac Newton the force of gravity on the particle F is given by;-

$$F = GMm/r^2 \qquad \qquad \text{... 22}$$

where G is Newton's Gravitational constant, M is the mass of the planet, m is the mass of the particle and r is the distance of the particle from the centre of the planet.

The Potential Energy for the particle is defined as;-

$$PE = \int F \, dr \qquad \text{... 23}$$

Therefore from equations 22 & 23 we can say;-

$$PE = [-GMm/r] \, ^{r=R_0}_{r=\infty} \qquad \text{... 24}$$

Where $R_0$ is the radius of the planet.

Therefore we can say;-

$$PE = - GMm/R_0 \qquad \text{...25}$$

Since the change in Potential Energy must match the change in Kinetic Energy for energy to be conserved then at speeds much below the speed of light we can ignore the effects of General Relativity and ignoring the effects of air resistance and applying Classical Physics we can say;-

$$1/2 \, mv^2 = GMm/R_0 \qquad \text{... 26}$$

or $\qquad v^2 = 2GM/R_0 \qquad \text{... 27}$

Putting in the values for the Mass of the Earth $M = 5.978 \times 10^{24}$ and the Radius of the Earth $R_0 = 6.371 \times 10^6$ we can deduce that the velocity of a particle reaching Earth that has gained all its speed from the fall, could be as high as;-

$$v = 11{,}190.76 \text{ m/s} \qquad \text{...28}$$

Any charged particle with a charge q that falls to Earth is going to be subject to the Earth's magnetic field B that flows from the Earth's magnetic North pole to the South pole. According to Lenz and

Faraday's law the resultant force F that is caused by this magnetic field on the moving charged particle moving at velocity v at right angles to the magnetic field would be;-

$$F = Bqv \qquad \qquad ...29$$

This force would cause the particle to move laterally which in turn would cause the resultant force to rotate slightly as it would always be at right angles both to the direction of the particle and the magnetic field. The resultant path described by the particle would therefore be a circle.

At low speeds we can ignore the effects of General Relativity and say for a circle;-

$$F = mv^2/r \qquad \qquad ...30$$

Using equations 29 & 30 we can therefore say;-

$$r = mv/Bq \qquad \qquad ... 31$$

where r is the radius of the circle described by the particle of mass m and charge q moving in magnetic field B.

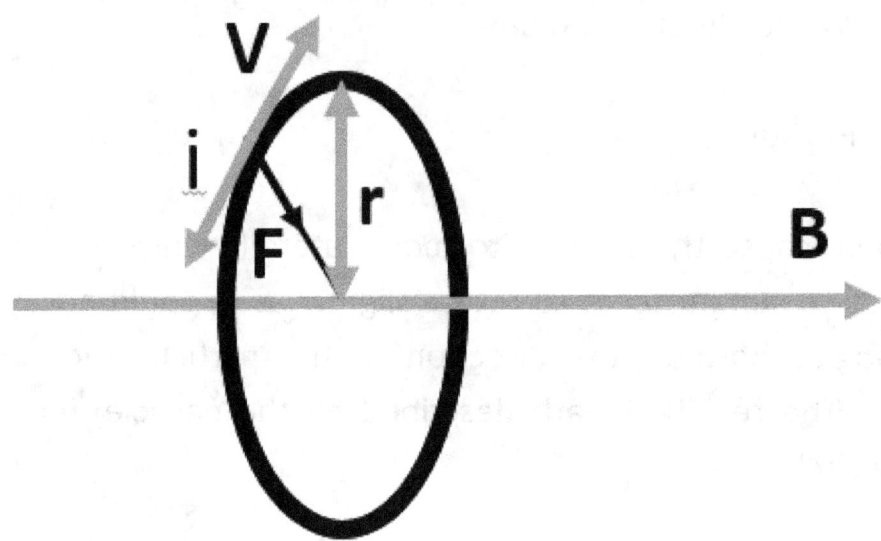

Diagram 4 The electrical current caused by the moving charged particle flows in the opposite direction to the motion for a negatively charged particle.

The maximum magnetic field strength of the Earth occurs at the poles and is;-

$$B = 6.5 \times 10^{-5} \text{ Tesla} \qquad \text{... 32}$$

If this magnetic field were in parallel to the surface of the Earth (which it *isn't* at the poles but we will assume it is just to get a limiting maximum value) we can say the minimum radius of the circular path described by a neutrino falling to Earth would be;-

$$r_{min} = 0.979 \times 10^{-3} \text{ m} \qquad \text{... 33}$$

The minimum magnetic field strength of the Earth occurs at the equator and is;-

$$B = 2.5 \times 10^{-5} \text{ Tesla} \qquad \qquad ... 34$$

The magnetic field at the equator is in parallel to the surface of the Earth so we can say the maximum radius of the circular path described by a neutrino falling to Earth would be;-

$$r_{max} = 2.545 \times 10^{-3} \text{ m} \qquad \qquad ... 35$$

The frequency at which a particle moves around a circular path is given by;-

$$f = v/2\pi r \qquad \qquad ... 36$$

So using the results from equations 28, 33 and 35 we can say that neutrinos in the Earth's magnetic field will rotate in circular paths close to the Earth's surface at frequencies f between;-

$$fmin = 0.6998 \text{ MHz} \qquad \qquad ...37$$
and
$$fmax = 1.8194 \text{ MHz} \qquad \qquad ...38$$

It should be possible to detect these rotations which are around the medium radio frequency range. Also if we were to increase the magnetic field strength B for example by using a permanent magnet, then we should be able to change the frequency of rotation of the neutrinos. This might prove to be an interesting opportunity as a research project for someone to pursue. Certainly there is significant "background" noise that is readily detectable by radio receivers approximately in this frequency range. Whether it can be attributed to the movement of neutrinos is open to question and can only be proven one way or another by experiment.

In theory this proposed change in frequency of rotation of the neutrinos, caused by the change in the Earth's magnetic field strength in moving from the poles to the equator, might be of use in measuring the latitude of any vehicle travelling on long journeys across the surface of the Earth either by land, sea or air but this remains to be confirmed.

All of this is speculative work but hopefully these estimations and approximate calculations may prove useful in helping to design experiments that will ultimately determine the exact nature of neutrinos. Despite their very small size, neutrinos are the most common particle in our Universe, and as such it is important that we endeavour to understand as much about them as possible, as they may well hold the key to unlocking further discoveries about the way our Universe has evolved, and how it may develop in the years ahead.

## Appendix 1 - The Relationship Between KE And PE

The Centripetal force on an electron travelling in a circular orbit according to General Relativity is given by;-

$$F = Y^N \, mv^2/r \qquad \text{...(A1.1)}$$

According to Coulomb the force between an electron and P protons can be expressed as;-

$$F = Y^N \times Pe^2/4\pi\varepsilon r^2 \qquad \text{...(A1.2)}$$

A1.1 & A1.2 =>

$$r = Pe^2/4\pi\varepsilon mv^2 \qquad \text{...(A1.3)}$$

by definition;-

$$Y = c/(c^2 - V^2)^{0.5} \qquad \text{...(A1.4)}$$

A1.3 & A1.4 =>

$$r = Pe^2 \times Y^2/4\pi\varepsilon mc^2(Y^2 - 1) \qquad \text{...(A1.5)}$$

A1.5 =>

$$dr/dY = -2Y \times Pe^2/\{4\pi\varepsilon mc^2 \, (Y^2 - 1)^2\} \qquad \text{...(A1.6)}$$

By definition Potential Energy PE is defined as;-

$$PE = \int F \, dr \qquad \text{...(A1.7)}$$

Equations A1.1, A1.5, A1.6 & A1.7 =>

$$PE = -2mc^2 \int Y^{(N-3)} \, dY \qquad \text{...(A1.8)}$$

Patrick Naughton

According to the theory of General Relativity N = 2

Therefore;-

**PE = -2mc² log$_e$ Y**                                    ...(A1.9)

This is derived independent of the spin S of the electron.

According to Einstein the Kinetic Energy of a particle is given by;

KE = mc²(Y - 1)                                    ...(A1.10)

If we plot -PE/KE for different values of Y we get the graph shown below;-

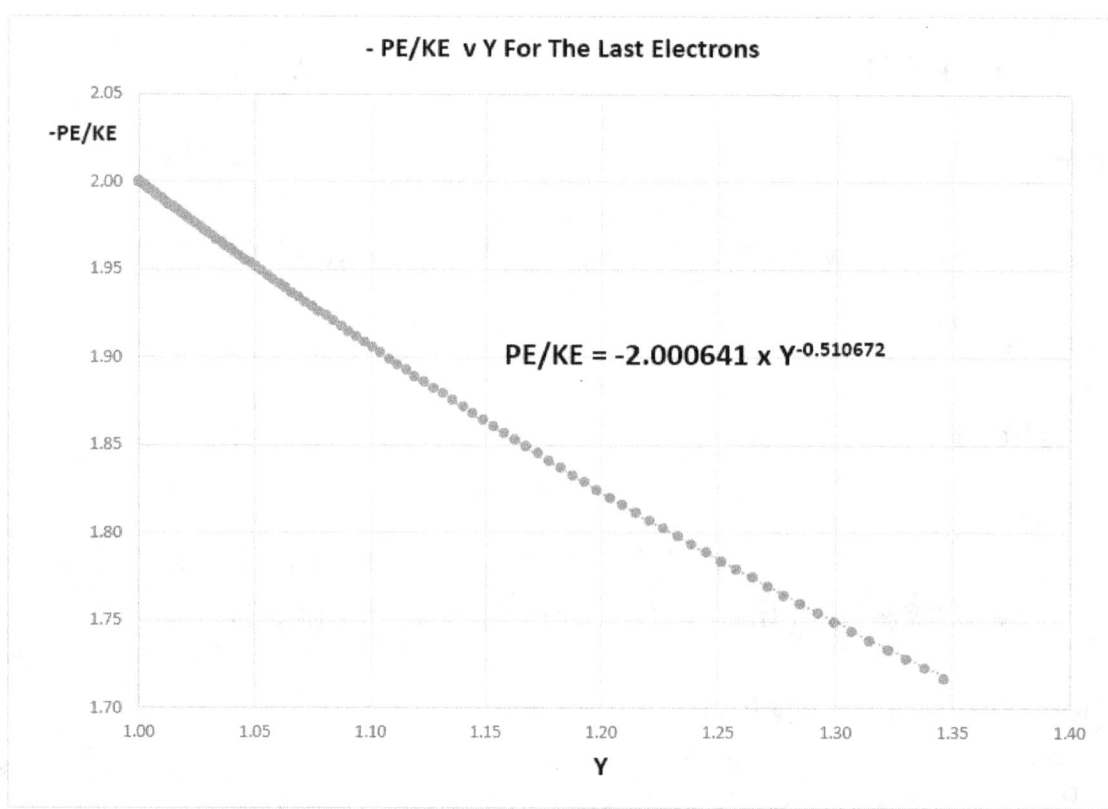

Graph A1

It can be seen that at low values of Y approaching 1 (i.e. when the velocity of the particle is small) then the ratio PE/KE approaches the

30

value -2 (approximately) which is in agreement with the result from Classical Physics.

The proof that Classical Physics states that PE/KE = -2 for a particle moving in a circle is given below.

In Classical Physics the centripetal force required for a particle to move in a circle is given by;

$$F = mv^2/r \qquad ...(A1.11)$$

According to Coulomb's law the force on an electron moving in an atom with P protons assuming no other force is;-

$$F = Pe^2/4\pi\varepsilon r^2 \qquad ...(A1.12)$$

But Potential Energy is defined as;-

$$PE = \int F dr \qquad ...(A1.13)$$

Therefore from equations A1.12 and A1.13 we can say;-

$$PE = -Pe^2/4\pi\varepsilon r \qquad ...(A1.14)$$

In Classical Physics Kinetic Energy is defined as;-

$$KE = 1/2 \; mv^2 \qquad ...(A1.15)$$

From equation A1.11 and A1.12 we can say;-

$$r = Pe^2/4\pi\varepsilon v^2 \qquad ...(A1.16)$$

Therefore from equations A1.14 and A1.16 we can say;

$$PE = -mv^2 \qquad \text{...(A1.17)}$$

So we can deduce that in Classical Physics for a particle travelling in a circle that;

$$PE = -2\text{x KE} \qquad \text{...(A1.18)}$$

## Appendix 2 - Ionisation Energies Last Electrons

The ionisation energies Eion for the *last* electrons in each atom are listed in the table below.

P is the number of Protons.
V is the velocity of the electron in m/s.
Y is the Lorentz coefficient.
R is the effective radius of the atom in metres.

| Proton P | Element | Eion Ion Energy Measured /eV | Eion Ion Energy Measured /J | V | Y | R/m |
|---|---|---|---|---|---|---|
| 1 | Hydrogen | 13.60 | 2.179E-18 | 2.187E+06 | 1.00003 | 5.293E-11 |
| 2 | Helium | 54.42 | 8.720E-18 | 4.375E+06 | 1.00011 | 2.646E-11 |
| 3 | Lithium | 122.47 | 1.962E-17 | 6.562E+06 | 1.00024 | 1.763E-11 |
| 4 | Beryllium | 217.75 | 3.489E-17 | 8.749E+06 | 1.00043 | 1.322E-11 |
| 5 | Boron | 340.27 | 5.452E-17 | 1.094E+07 | 1.00067 | 1.058E-11 |
| 6 | Carbon | 490.06 | 7.852E-17 | 1.312E+07 | 1.00096 | 8.811E-12 |
| 7 | Nitrogen | 667.13 | 1.069E-16 | 1.530E+07 | 1.00131 | 7.550E-12 |
| 8 | Oxygen | 871.52 | 1.396E-16 | 1.749E+07 | 1.00171 | 6.603E-12 |
| 9 | Fluorine | 1,103.26 | 1.768E-16 | 1.967E+07 | 1.00216 | 5.867E-12 |
| 10 | Neon | 1,362.38 | 2.183E-16 | 2.185E+07 | 1.00267 | 5.278E-12 |
| 11 | Sodium | 1,648.92 | 2.642E-16 | 2.403E+07 | 1.00323 | 4.795E-12 |
| 12 | Magnesium | 1,962.92 | 3.145E-16 | 2.620E+07 | 1.00384 | 4.393E-12 |
| 13 | Aluminium | 2,304.44 | 3.692E-16 | 2.838E+07 | 1.00451 | 4.052E-12 |
| 14 | Silicon | 2,673.53 | 4.284E-16 | 3.055E+07 | 1.00523 | 3.760E-12 |
| 15 | Phosphorous | 3,070.24 | 4.919E-16 | 3.272E+07 | 1.00601 | 3.507E-12 |
| 16 | Sulphur | 3,494.65 | 5.599E-16 | 3.488E+07 | 1.00684 | 3.285E-12 |
| 17 | Chlorine | 3,946.81 | 6.324E-16 | 3.705E+07 | 1.00772 | 3.089E-12 |
| 18 | Argon | 4,426.81 | 7.093E-16 | 3.921E+07 | 1.00866 | 2.915E-12 |
| 19 | Potassium | 4,934.68 | 7.906E-16 | 4.136E+07 | 1.00966 | 2.759E-12 |
| 20 | Calcium | 5,470.58 | 8.765E-16 | 4.352E+07 | 1.01071 | 2.618E-12 |
| 21 | Scandium | 6,034.50 | 9.668E-16 | 4.567E+07 | 1.01181 | 2.491E-12 |
| 22 | Titanium | 6,625.82 | 1.062E-15 | 4.781E+07 | 1.01297 | 2.375E-12 |
| 23 | Vanadium | 7,246.12 | 1.161E-15 | 4.996E+07 | 1.01418 | 2.269E-12 |
| 24 | Chromium | 7,894.81 | 1.265E-15 | 5.210E+07 | 1.01545 | 2.172E-12 |
| 25 | Manganese | 8,571.94 | 1.373E-15 | 5.423E+07 | 1.01678 | 2.082E-12 |

| 26 | Iron | 9,277.69 | 1.486E-15 | 5.636E+07 | 1.01816 | 1.999E-12 |
|---|---|---|---|---|---|---|
| 27 | Cobalt | 10,012.12 | 1.604E-15 | 5.849E+07 | 1.01959 | 1.923E-12 |
| 28 | Nickel | 10,775.40 | 1.726E-15 | 6.061E+07 | 1.02109 | 1.851E-12 |
| 29 | Copper | 11,567.62 | 1.853E-15 | 6.273E+07 | 1.02264 | 1.785E-12 |
| 30 | Zinc | 12,388.93 | 1.985E-15 | 6.484E+07 | 1.02424 | 1.722E-12 |
| 31 | Gallium | 13,239.49 | 2.121E-15 | 6.695E+07 | 1.02591 | 1.664E-12 |
| 32 | Germanium | 14,119.43 | 2.262E-15 | 6.905E+07 | 1.02763 | 1.610E-12 |
| 33 | Arsenic | 15,028.91 | 2.408E-15 | 7.115E+07 | 1.02941 | 1.558E-12 |
| 34 | Selenium | 15,968.08 | 2.558E-15 | 7.324E+07 | 1.03125 | 1.510E-12 |
| 35 | Bromine | 16,937.13 | 2.714E-15 | 7.533E+07 | 1.03315 | 1.464E-12 |
| 36 | Krypton | 17,936.21 | 2.874E-15 | 7.741E+07 | 1.03510 | 1.420E-12 |
| 37 | Rubidium | 18,965.52 | 3.039E-15 | 7.948E+07 | 1.03711 | 1.379E-12 |
| 38 | Strontium | 20,025.23 | 3.208E-15 | 8.155E+07 | 1.03919 | 1.340E-12 |
| 39 | Yttrium | 21,115.55 | 3.383E-15 | 8.362E+07 | 1.04132 | 1.303E-12 |
| 40 | Zirconium | 22,236.68 | 3.563E-15 | 8.567E+07 | 1.04352 | 1.268E-12 |
| 41 | Niobium | 23,388.80 | 3.747E-15 | 8.772E+07 | 1.04577 | 1.234E-12 |
| 42 | Molybdenum | 24,572.15 | 3.937E-15 | 8.977E+07 | 1.04809 | 1.202E-12 |
| 43 | Technetium | 25,786.99 | 4.132E-15 | 9.180E+07 | 1.05046 | 1.172E-12 |
| 44 | Ruthenium | 27,033.50 | 4.331E-15 | 9.383E+07 | 1.05290 | 1.142E-12 |
| 45 | Rhodium | 28,311.96 | 4.536E-15 | 9.586E+07 | 1.05541 | 1.114E-12 |
| 46 | Palladium | 29,622.60 | 4.746E-15 | 9.787E+07 | 1.05797 | 1.087E-12 |
| 47 | Silver | 30,965.70 | 4.961E-15 | 9.988E+07 | 1.06060 | 1.062E-12 |
| 48 | Cadmium | 32,341.49 | 5.182E-15 | 1.019E+08 | 1.06329 | 1.037E-12 |
| 49 | Indium | 33,750.31 | 5.407E-15 | 1.039E+08 | 1.06605 | 1.013E-12 |
| 50 | Tin | 35,192.39 | 5.638E-15 | 1.059E+08 | 1.06887 | 9.901E-13 |
| 51 | Antimony | 36,668.05 | 5.875E-15 | 1.079E+08 | 1.07176 | 9.680E-13 |
| 52 | Tellurium | 38,177.56 | 6.117E-15 | 1.098E+08 | 1.07471 | 9.467E-13 |
| 53 | Iodine | 39,721.41 | 6.364E-15 | 1.118E+08 | 1.07773 | 9.262E-13 |
| 54 | Xenon | 41,299.71 | 6.617E-15 | 1.137E+08 | 1.08082 | 9.064E-13 |
| 55 | Caesium | 42,912.99 | 6.875E-15 | 1.157E+08 | 1.08398 | 8.873E-13 |
| 56 | Barium | 44,561.47 | 7.140E-15 | 1.176E+08 | 1.08721 | 8.688E-13 |
| 57 | Lanthanum | 46,245.60 | 7.409E-15 | 1.196E+08 | 1.09050 | 8.510E-13 |
| 58 | Cerium | 47,965.72 | 7.685E-15 | 1.215E+08 | 1.09387 | 8.337E-13 |
| 59 | Praesodymium | 49,722.25 | 7.966E-15 | 1.234E+08 | 1.09730 | 8.169E-13 |
| 60 | Neodymium | 51,515.58 | 8.254E-15 | 1.253E+08 | 1.10081 | 8.007E-13 |
| 61 | Promethium | 53,346.10 | 8.547E-15 | 1.272E+08 | 1.10440 | 7.849E-13 |
| 62 | Samarium | 55,214.23 | 8.846E-15 | 1.291E+08 | 1.10805 | 7.697E-13 |
| 63 | Europium | 57,120.63 | 9.152E-15 | 1.310E+08 | 1.11178 | 7.548E-13 |
| 64 | Gadolinium | 59,065.52 | 9.463E-15 | 1.329E+08 | 1.11559 | 7.404E-13 |
| 65 | Terbium | 61,049.64 | 9.781E-15 | 1.348E+08 | 1.11947 | 7.264E-13 |

| 66 | Dysprosium | 63,073.50 | 1.011E-14 | 1.366E+08 | 1.12343 | 7.128E-13 |
|----|------------|-----------|-----------|-----------|---------|-----------|
| 67 | Holmium | 65,136.80 | 1.044E-14 | 1.385E+08 | 1.12747 | 6.996E-13 |
| 68 | Erbium | 67,241.80 | 1.077E-14 | 1.403E+08 | 1.13159 | 6.867E-13 |
| 69 | Thulium | 69,387.30 | 1.112E-14 | 1.421E+08 | 1.13579 | 6.742E-13 |
| 70 | Ytterbium | 71,574.80 | 1.147E-14 | 1.440E+08 | 1.14007 | 6.620E-13 |
| 71 | Lutetium | 73,804.80 | 1.182E-14 | 1.458E+08 | 1.14443 | 6.501E-13 |
| 72 | Hafnium | 76,077.80 | 1.219E-14 | 1.476E+08 | 1.14888 | 6.385E-13 |
| 73 | Tantalum | 78,394.70 | 1.256E-14 | 1.494E+08 | 1.15342 | 6.272E-13 |
| 74 | Tungsten | 80,755.60 | 1.294E-14 | 1.512E+08 | 1.15804 | 6.161E-13 |
| 75 | Rhenium | 83,162.30 | 1.332E-14 | 1.530E+08 | 1.16275 | 6.053E-13 |
| 76 | Osmium | 85,614.40 | 1.372E-14 | 1.547E+08 | 1.16754 | 5.948E-13 |
| 77 | Iridium | 88,113.30 | 1.412E-14 | 1.565E+08 | 1.17244 | 5.845E-13 |
| 78 | Platinum | 90,659.70 | 1.453E-14 | 1.583E+08 | 1.17742 | 5.745E-13 |
| 79 | Gold | 93,254.30 | 1.494E-14 | 1.600E+08 | 1.18250 | 5.647E-13 |
| 80 | Mercury | 95,897.70 | 1.536E-14 | 1.617E+08 | 1.18767 | 5.551E-13 |
| 81 | Thallium | 98,591.60 | 1.580E-14 | 1.635E+08 | 1.19294 | 5.457E-13 |
| 82 | Lead | 101,336.40 | 1.624E-14 | 1.652E+08 | 1.19831 | 5.365E-13 |
| 83 | Bismuth | 104,132.80 | 1.668E-14 | 1.669E+08 | 1.20378 | 5.275E-13 |
| 84 | Polonium | 106,982.70 | 1.714E-14 | 1.686E+08 | 1.20936 | 5.186E-13 |
| 85 | Astatine | 109,886.00 | 1.761E-14 | 1.703E+08 | 1.21504 | 5.100E-13 |
| 86 | Radon | 112,843.70 | 1.808E-14 | 1.720E+08 | 1.22083 | 5.015E-13 |
| 87 | Francium | 115,859.00 | 1.856E-14 | 1.736E+08 | 1.22673 | 4.932E-13 |
| 88 | Radium | 118,931.30 | 1.906E-14 | 1.753E+08 | 1.23275 | 4.851E-13 |
| 89 | Actinium | 122,062.90 | 1.956E-14 | 1.770E+08 | 1.23887 | 4.771E-13 |
| 90 | Thorium | 125,253.40 | 2.007E-14 | 1.786E+08 | 1.24512 | 4.693E-13 |
| 91 | Protoactinium | 128,507.10 | 2.059E-14 | 1.803E+08 | 1.25148 | 4.616E-13 |
| 92 | Uranium | 131,821.20 | 2.112E-14 | 1.819E+08 | 1.25797 | 4.541E-13 |
| 93 | Neptunium | 135,202.40 | 2.166E-14 | 1.835E+08 | 1.26459 | 4.467E-13 |
| 94 | Plutonium | 138,646.00 | 2.221E-14 | 1.851E+08 | 1.27133 | 4.394E-13 |
| 95 | Americium | 142,161.00 | 2.278E-14 | 1.867E+08 | 1.27820 | 4.323E-13 |
| 96 | Curium | 145,743.00 | 2.335E-14 | 1.883E+08 | 1.28521 | 4.252E-13 |
| 97 | Berkelium | 149,398.00 | 2.394E-14 | 1.899E+08 | 1.29237 | 4.183E-13 |
| 98 | Californium | 153,124.00 | 2.453E-14 | 1.915E+08 | 1.29966 | 4.116E-13 |
| 99 | Einsteinium | 156,926.00 | 2.514E-14 | 1.931E+08 | 1.30710 | 4.049E-13 |
| 100 | Fermium | 160,804.00 | 2.576E-14 | 1.946E+08 | 1.31469 | 3.983E-13 |
| 101 | Mendelevium | 164,764.00 | 2.640E-14 | 1.962E+08 | 1.32244 | 3.918E-13 |
| 102 | Nobelium | 168,806.00 | 2.705E-14 | 1.977E+08 | 1.33035 | 3.855E-13 |
| 103 | Lawrencium | 172,930.00 | 2.771E-14 | 1.993E+08 | 1.33842 | 3.792E-13 |
| 104 | Rutherfordium | 177,148.00 | 2.838E-14 | 2.008E+08 | 1.34667 | 3.730E-13 |

## References

1. DK Space Encyclopaedia - Heather Couper & Nigel Henbest p154 DK Publications c2008.

/ends

# Neutron Charge

Neutron Charge

Patrick Naughton

# Neutron Charge

Patrick Naughton

# Neutron Charge

www.ingramcontent.com/pod-product-compliance
Lightning Source LLC
Chambersburg PA
CBHW081307180526
45170CB00007B/2601